HAMSHACK
RASPBERRY PI

HOW TO USE THE RASPBERRY PI
FOR AMATEUR RADIO ACTIVITIES

JAMES BAUGHN K9EOH

authorHOUSE®

AuthorHouse™
1663 Liberty Drive
Bloomington, IN 47403
www.authorhouse.com
Phone: 1 (800) 839-8640

Published by AuthorHouse 05/12/2017

ISBN: 978-1-5246-9165-3 (sc)
ISBN: 978-1-5246-9163-9 (hc)
ISBN: 978-1-5246-9164-6 (e)

Library of Congress Control Number: 2017907546

Print information available on the last page.

This book is printed on acid-free paper.

James Baughn made his first ham radio contact as KN9EOH, operating Morse code on September 23, 1956. He passed his General license in Chicago and while in the Navy, as a radioman, he earned his Extra Class License. He was active in the Navy Military Affiliate Radio System relaying messages to and from military personnel in Vietnam and their loved ones in America. Living in "Sweet" Owen County Indiana he became heavily involved with the amateur radio community. He was the Owen County Amateur Radio Association's public information officer, secretary and webmaster for their site www.owencountyara.org and has been the Owen County Emergency Coordinator. He checks into the Indiana Traffic Net and Indiana Digital Traffic net.

He worked with computer hardware and software beginning in 1966 when he joined IBM as a Field Engineer. He opened one of the first home computer stores in Indianapolis in early 1977. Later he covered Indiana and surrounding states working on DEC, Data General and Apollo CAD/CAM systems. He retired from Eli Lilly pharmaceutical company's Information Technology division in 2007.

He believes the rise of the Raspberry Pi today is as exciting as the rise of the personal computers in the '70s.

TABLE OF CONTENTS

Chapter 1 Introduction to the Raspberry Pi 1

    Accessories for the Raspberry Pi....................................3
    A short development history 4
    Program installation sections 5
    Task Bar Information..6

Chapter 2 Setup 9

    Wi-Fi setup 12
    SD Card Configuration.. 13
    Backing up the SD card...22
    Turning the Raspberry Pi on and off 22

Chapter 3 Initial Software 23

    Samba Installation 25
    Printer Setup[3] 27
    Adding Printers 29

Chapter 4 Adding Ham Radio Programs...........................31

    Hamradiomenus: Establishes Menus...........................33
    Aldo: Morse Code 35
    Chirp: Configuration tool for amateur radios 38
    Fldigi: Digital Modem Program 41
    gpredict: Satellite Tracking 51

Trusted QSL-Logbook of the World 56
Gworldclock - World Clock 59
Xlog: a logging program for Hams 62
Xnec2c: Calculate and Display Antenna Properties.....67

Appendix A Current Pi Model Information........................71

Appendix B CLI Useful commands.....................................77

Appendix C Useful websites 81

Appendix D Bibliography..85

Appendix E Glossary...89

Appendix F End Notes...95

CHAPTER 1

Introduction to the Raspberry Pi

Accessories for the Raspberry Pi

In order to install the programs in this book the minimum requirements are:

1) Raspberry Pi any model, although the Raspberry Pi 3 Model B is recommended.

2) A 5V 2.4A Switching Power Supply with Micro USB cable.

3) If using other than the Raspberry 3 Model B, a Wi-Fi USB fob would be needed to access a Wi-Fi system.

4) Case for the Raspberry Pi.

The above items can be purchased from any of the following websites; www.adafruit.com www.amazon.com www. element14.com www.thepihut.com.

5) USB Keyboard and Mouse or a wireless USB Keyboard and Mouse.

6) HDMI TV for a Monitor.

There would also be some specific accessories for those programs that connect with Computer Aided Transceiver (CAT) radios.

1) SignalLink USB audio interface www.tigertronics.com

2) If your CAT radio connector is RS-232, a USB to 9 pin serial converter would be required. www.amazon.com

3) Interface cable for a Computer Assisted Transceiver (CAT). Check with any amateur radio supply house for you transceiver's matching cable.

A short development history

In 1971 Intel produced the first general purpose programmable microprocessor chip, the 4004. It was one of four chips developed for a printing calculator. It delivered the processing power of the first computer built in 1946 which filled an entire room. This chip was followed by the 8008 and then the 8080. The latter was used in the Altair which became the first successfully marketed computer kit and set standards used for several years.

During the beginning of this personal computer 'revolution' the machines required more computer knowledge to operate than most people possessed. In the late 70s computer stores began to open but only customers with some computer background could successfully use them, as few well working programs were available. Just to put a program into the computer required using switches on the front panel is specific order. However these early computers attracted many young people who went on to college computer courses.

Over time the personal computer became an expensive household appliance. Parents were reluctant to allow their children to use the family computers because of the cost of the machines and the possibility of the children erasing the family data.

In 2006 Dr. Eben Upton and associates of the University of Cambridge Computer Laboratory noted the declining numbers and skills of students applying for computer science courses. It was decided to develop an affordable computer that young people could use to become familiar with computer concepts.

In 2011 the Raspberry Pi Model B was born and sold over two million units within two years. Since then there has been continuing improvements in various models of the Raspberry Pi. The machines are not only computers but they are also microcontrollers with pins that can externally sense and control devices.

Some of the uses, of these computers, are general purpose computing, learning to program, a project platform, prototyping products, creating a media center, control of robots, home automation and security systems.

The emphasis of this book is to provide guidance in setting up and operating a Raspberry Pi providing various amateur radio programs.

Program installation sections

Each program installation section will consist of the following parts:

Overview - a description of the program to be installed.

Hardware: any additional hardware required.

Installation: Steps to install the software. In most cases there are two methods to do so. Each method will be explained in the program Installation sections.

Desktop Entry File: The Desktop Entry file configures how an application appears in the desktop menu. Those programs that do not include a Desktop Entry file will require one be created.

Configuration: How to initially configure the program.

Operation: Operating the program.

Task Bar Information

From time to time there will be references to the "Task Bar" located at the top of the screen.

Task Bar

The first icon in the taskbar is the Menu selection icon. The second is the Internet browser. The third is the file browser. The fourth is the Command Line Interface Terminal. The fifth is the Mathematica program. The sixth is the Wolfram program.

See http://www.i-programmer.info/news/91-hardware/ 7620-mathematica-10-now-on-raspberry-pi.html for more information about Mathematica and Wolfram.

The next icon is the blue tooth control icon.

See https://www.cnet.com/how-to/how-to-setup-blue tooth-on-a-raspberry-pi-3/ for more information.

The next icon indicates Wi-Fi is attached. If the Wi-Fi is not connected, that icon will look like this:

Wi-Fi Off Icon

The following icons are the speaker volume control, the CPU memory in use percentage, the time and the last is the eject drive icon.

CHAPTER 2

Setup

New Out Of Box Software (NOOBS) is a program used to easily install the operating system on the Raspberry Pi Secure Digital (SD) card.

The Raspberry Pi operating system, Raspbian, and data storage are kept on a Micro SD card. Thus you can set up various SD cards each booting a Raspberry Pi in different configurations. For example, by changing the SD Card in the Pi, the Pi could be a SETI[1] cruncher, robot, drone control system, media system, home security system, camera controller, GPS, earthquake detector, RFID Reader, Weather station, Radon detector, etc.[2]

The speed of micro SD cards range up to class 10 which is the fastest. The class is indicated by a number in a circle. For the Raspberry Pi it is recommended that the minimum useful class is 4. The class 10 card will also operate for a longer period of time. However, in time the card will "wear out." It is highly recommended, as with any computer data, that one backs up the SD card. See "Backing up the SD Card."

If your Raspberry Pi came with a pre-installed NOOBS SD card, skip to 'Wi-Fi setup.'

Download NOOBS from https://www.raspberrypi.org/downloads/

Click on the NOOBS icon

Under the NOOBS icon select [Download Zip]

Unzip the Downloaded NOOBS system folder.

Follow the INSTRUCTIONS-README.txt file found in the unzipped NOOBS folder.

On the bottom of the desktop is the language selection window. Select the language for your country. For example, in the US select English (US). The corresponding US keyboard should be displayed.

Click on the Raspbian check box and click on the install icon. Click [yes] on the Confirm window. When the [OS(es) Installed] window appears, click on OK.

Wi-Fi setup

Wi-Fi off icon

In order to download the programs in this book, you must be connected to the Internet. Therefore, you have two options. You can connect to your internet router with a Ethernet cable or attach to your Wi-fi router. This section explains how to connect to a Wi-fi router.

Connection requires a Wi-Fi USB fob if you using a Raspberry Pi other than the Raspberry Pi 3 Model B. Configure the connection by clicking on the icon displaying two vertical lines on the right side of the task bar. Available Wi-Fi spots will be displayed. Click on the desired Wi-Fi spot connection enter the Wi-Fi network password and the connection should be made.

After connecting to the Wi-Fi and in order to update to the latest version of the Raspbian operating system, open a terminal session by clicking on the terminal icon on the task bar at the top of the desktop.

If not recently done update information about new and updated Linux packages that are available on the Internet:

$ sudo apt-get update

If a new version of a package installed on your machine is available the package will be upgraded with the following command. No installed packages are ever removed by this command.

$ sudo apt-get upgrade

Press the [Enter] key in response to the 'Do You Want to Continue [Y/n]' prompt.

In order to update the Raspberry Pi firmware:

$ sudo apt-get install rpi-update

SD Card Configuration

In order to configure the SD card, click on the task bar raspberry icon. Then select [Preferences] and [Raspberry Configuration] [System].

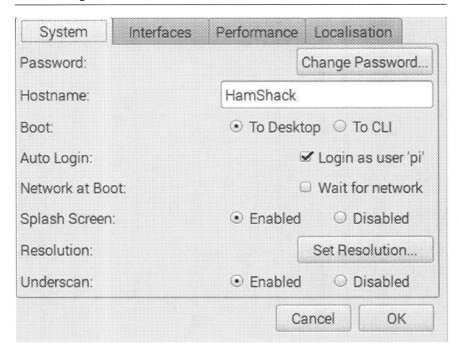

Raspberry Pi Configuration System Tab

Password: It is highly recommended that you change the Password. The default password for the Raspberry Pi is raspberry. If you have not changed it, enter that into the Current password box then enter you new password in the two other boxes and click on the [OK] button. If the [Ok] button is grayed out, the two entries of the password you wish to use, do not match.

Hostname: You can also change the host-name to identify the Raspberry Pi on network listings. I have changed the Host-name to Hamshack.

EPSON   HAMSHACK   JAMES-PC   JAMES-   K9EOH
WF-3640   UBUNTU
Series

Network Hosts

This way one can easily determine which computers are being seen on the network. In the above network list, Epson WF-3640 Series is the network printer, HAMSHACK is the system we are setting up, JAMES PC is a Windows laptop, JAMES UBUNTU is a Linux desktop and K9EOH is another Raspberry Pi.

Boot: You can choose for the Pi to boot to a Graphical User Interface (Desktop) or Command Line Interface (CLI).

The Graphical User Interface (GUI) desktop

The Graphical User Interface uses windows, icons and menus which can be manipulated by a mouse.

The Command User Interface (CLI)

All interaction with the Raspberry Pi using CLI is accomplished by text input from the keyboard that is converted to the appropriate operating functions.

Auto login: if checked the system will boot directly into the Boot configuration selected above. If it is not checked, the password will be required to continue booting.

Network at Boot: If "Wait for network" is checked the system will pause during boot up to wait for a network connection to be made.

Splash Screen: If enabled a splash screen "Welcome to PIXEL" will be displayed during boot.

Resolution: Resolution for the monitor can be set.

Underscan: If the screen does not 'fit' the monitor, change to Enable or Disable to correct the problem.

Then select the [Interfaces] tab.

System	Interfaces	Performance	Localisation
Camera:		○ Enabled	⦿ Disabled
SSH:		○ Enabled	⦿ Disabled
VNC:		○ Enabled	⦿ Disabled
SPI:		○ Enabled	⦿ Disabled
I2C:		○ Enabled	⦿ Disabled
Serial:		○ Enabled	⦿ Disabled
1-Wire:		○ Enabled	⦿ Disabled
Remote GPIO:		○ Enabled	⦿ Disabled

Cancel   OK

Raspberry Pi Configuration Interfaces Tab

Camera: If a Raspberry Pi camera is installed click on the Enable button

SSH: With Secure Shell (SSH) enabled, a cryptographic network protocol for operating securely over an unsecured network is used with certain programs. This would be needed to allow remote operation of the Raspberry Pi. See "Headless Operating".

VNC: Virtual Network Computing is a method to securely

operate a computer over the network by displaying the desktop of the remote system and allowing changes to be made by mouse and keyboard by the local machine.

SPI: Serial Peripheral Interface is a synchronous serial communication interface specification for short distance communication. See https://en.wikipedia.org/wiki/Serial_Peripheral_Interface_Bus

I2C: Inter IC allows easy communication between components on the same circuit board or linked to components via cable. See https://www.i2c-bus.org/i2c-bus/

Serial: A Universal Asynchronous Receiver/Transmitter (UART) port for a serial console. See http://www.raspberry-projects.com/pi/programming-in-c/uart-serial-port/using-the-uart

1-Wire: provides low-speed data, power and signaling with a single conductor and ground. See https://www.maximintegrated.com/en/products/digital/one-wire.html

Remote GPIO: Server which allows remote network computers to access the state of General Purpose Input Output (GPIO) pins. See an example of use at http://www.instructables.com/id/Remote-control-Raspberry-PI-GPIO-pins-over-the-int/

Select the [Performance] tab.

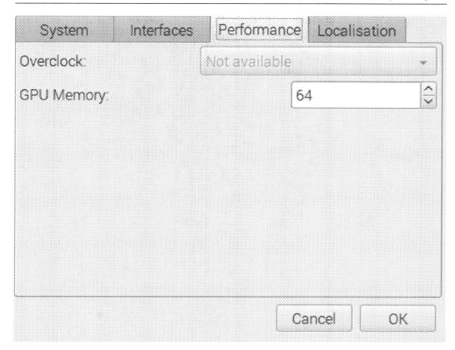

Raspberry Pi Configuration Performance Tab

Overclock: Previous models of Raspberry Pi had the ability to increase the speed of the CPU clock. However the increase in speed could result in instability of the system and even make the SD card unbootable. The Raspberry Pi 3 Model B does not allow over clocking.

GPU Memory: The amount of memory used by the Graphics Processing Unit is subtracted from the Central Processing Unit (CPU). By increasing the GPU memory, graphics can run faster.

Select the [Localisation] tab.

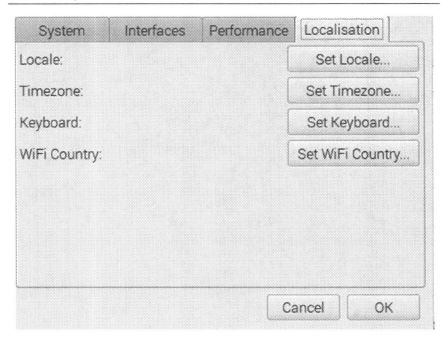

Raspberry Pi Configuration Localisation Tab

Set Local: Set the language, Country and Character Set for your location.

Set Timezone: Set your time zone.

Set Keyboard: Set the type of keyboard you are using.

Set Wi-Fi Country: Countries have different operating frequencies allotted to Wi-Fi usage. Set the country in which you are operating.

When finished, click on [OK].

In the [Reboot needed] window click on the [Yes] button.

The system will reboot and all selections will be saved.

NOTE: If, when the system reboots after the configuration section and you are unable to see the raspberry menu icon, click on the Terminal icon and enter:

$ sudo raspi-config

You will see the Raspberry Pi Software Configuration Tool.

Down arrow to selection "Advanced Options" and press the [Enter} key.

Then tab to the "Overscan" selection and press [Enter].

Tab to <yes> and press [Enter] twice.

Tab to <finish> and press [Enter].

In the Terminal window type:

$ sudo reboot –h now

The system will reboot and one should be able to see the complete task bar, including the raspberry icon, at the top of the screen.

END OF NOTE

After rebooting, you should observe a desktop or command-line interface as selected in the Raspberry Pi Configuration System tab SD Card Configuration section above.

Backing up the SD card

After you have put applications on your card, don't forget to back it up. See the website http://raspi.tv/2012/back-up-your-raspberry-pi-sd-card for instructions for Windows and Linux.

Turning the Raspberry Pi on and off

Because the Raspberry Pi has no on/off switch, select [Menu] [Shutdown] and click on [OK] on the End Session window.

If you intend to make any changes to the raspberry hardware after shutting down unplug the power connector.

To restart the machine after shutting it down, unplug the power connector and plug it back in.

CHAPTER 2

Initial Software

Samba Installation

Overview: You can share files and folders with other computers on your local network by installing Samba. For more information about samba, go to https://www.samba.org/.

Hardware: No additional hardware is required.

Installation by command line:

Open a terminal session by clicking on the terminal icon on the task bar at the top of the screen.

If not recently done update information about new and updated Linux packages that are available:

$ sudo apt-get update

If a new version of a package installed on your machine is available the package will be upgraded with the following command. No installed packages are ever removed by this command.

$ sudo apt-get upgrade

Then in order to install the program:

$ sudo apt-get install samba samba-common-bin

Press the [Enter] key in response to the 'Do You Want to Continue [Y/n]' prompt.

Configuration:

When the program has been loaded:

$ sudo nano /etc/samba/smb.conf

Scroll down to Global Settings and ensure workgroup = [Your Network Workgroup Name] and remove the # from the line [# wins support = no].

```
[global]

### Browsing/Identification ###

# Change this to the workgroup/NT-domain name your Samba server will $
    workgroup = WORKGROUP

# Windows Internet Name Serving Support Section:
# WINS Support - Tells the NMBD component of Samba to enable its WINS$
    wins support = no
```

Smb Edit Window

Scroll down to the end and add the following, as shown below, to create a directory that others on the network can access.

```
[pi home]
        comment=HAMSHACK
        path=/home/pi
        browseable=Yes
        only guest=no
        create mask=0777
        directory mask=0777
        public=no
```

[Pi Home] edit

To save the changes type "[Ctrl]X", "Y", [Enter]

Test the smb.conf file

$ sudo testparm /etc/samba/smb.conf

To restart Samba Enter:

$ sudo /etc/init.d/smbd restart

Printer Setup[3]

Overview:

The program CUPS connects printers to the computer.

Hardware:

A printer to be attached

Installation by command line:

Open a terminal session by clicking on the terminal icon on the task bar at the top of the screen.

If not recently done update information about new and updated Linux packages that are available:

$ sudo apt-get update

If a new version of a package installed on your machine is available the package will be upgraded with the following command. No installed packages are ever removed by this command.

$ sudo apt-get upgrade

Then in order to install the program:

$ sudo apt-get install cups

Press enter in response to the 'Do You Want to Continue [Y/n]' prompt.

Installation by Add/Remove Software menu selection:

Click on the Raspberry Icon on the task bar.

Click on preferences on the drop down list.

Select Add/Remove Software.

Enter cups in the search box and press enter.

Click on the cups menus for "Common UNIX Printing System™ - PPD/driver support, web interface".

Click on the Apply button.

Enter the pi password.

When the download is complete you have completed the setup.

After completion of either method of download:

Add the Pi user ID to the user group that has access to the printers and printer queue:

Open a terminal session by clicking on the terminal icon in the task bar and enter:

$ sudo usermod –a –G lpadmin pi

After completing the above commands restart cups:

$ sudo /etc/init.d/cups restart

Adding Printers

To add a printer open a browser on the Raspberry Pi with a URL of "localhost:631".

Click on the [Administration] tab and [Add Printer] button. Enter username: pi password: raspberry (or password to which you have changed it in the configuration menu).

Select the printer to be added and click [continue]. Your selected printer should be listed, click [Continue].

Select your printer driver in the list and click [Continue].

Review the Default options and click [Set Default Options]

The Job Status page will then be displayed.

The CUPS page can then be closed.

CHAPTER 4

Adding Ham Radio Programs

This chapter will discuss the installation of useful Linux amateur radio software. A list of such software can be found at http://www.raspberryconnect.com/raspbian-packages-list/item/71-raspbian-hamradio.

Hamradiomenus: Establishes Menus

Overview:

This program will create a sub-menu for ham radio applications that have been installed. It will only appear when at least one application has been installed with a Desktop Entry file containing the category entry "HamRadio".

Hardware:

There are no additional hardware requirements.

Installation:

There are two ways to install software. Choose one of the following methods:

Installation by command line:

Open a terminal session by clicking on the terminal icon on the task bar at the top of the screen.

If not recently done update information about new and updated Linux packages that are available:

$ sudo apt-get update

If a new version of a package installed on your machine is available the package will be upgraded with the following command. No installed packages are ever removed by this command.

$ sudo apt-get upgrade

Press the [Enter] key in response to the 'Do You Want to Continue [Y/n]' prompt.

Then in order to install the program:

$ sudo apt-get install hamradiomenus

In response to the 'Do You Want to Continue [Y/n]' prompt, press the [Enter] key.

When the download is complete you have completed the installation.

Installation by Add/Remove Software menu selection:

Click on the Raspberry Icon on the task bar.

Click on preferences on the drop down list.

Select Add/Remove Software.

Enter HamRadioMenus in the search box and press enter.

Click on the "hamradio menus for GNOME and KDE" selection square.

Click on the Apply button.

Enter the pi password.

When the download is complete you have completed the installation.

Desktop Entry:

Not required by hamradiomenus!

Configuration:

Configuration is not needed.

Operating:

When an application of the HamRadio category is installed, the hamradiomenus icon will appear.

Aldo: Morse Code

Overview:

Aldo is a Morse code learning tool providing four types of training methods displayed as the startup menu:

1: Blocks method - identify blocks of random characters played in Morse code.

2: Koch - two characters will be played at full speed until you're able to identify 90 percent of them. Then one more character is added and so on.

3: Read from file - send characters generated from a file.

4: Call Sign - identify random callsigns played in Morse code.

5: Settings: Set up speed and methods of choosing letters to be sent.

6: Exit program

Hardware:

There are no additional hardware requirements.

Installation:

There are two ways to install software. Choose one of the following methods:

Installation by command line:

Open a terminal session by clicking on the terminal icon on the task bar at the top of the screen.

If not recently done update information about new and updated Linux packages that are available:

$ sudo apt-get update

If a new version of a package installed on your machine is available the package will be upgraded with the following command. No installed packages are ever removed by this command.

$ sudo apt-get upgrade

Then in order to install the program:

$ sudo apt-get install aldo

Press the [Enter] key in response to the 'Do You Want to Continue [Y/n]' prompt.

When the download is complete you have completed the installation.

Installation by Add/Remove Software menu selection:

Click on the Raspberry Icon on the task bar.

Click on preferences on the drop down list.

Select Add/Remove Software.

Enter aldo in the search box and press enter.

Click on the "Morse Code training program" selection square.

Click on the Apply button.

Enter the pi password.

When the download is complete you have completed the installation.

Desktop Entry:

In order to create a start menu window for aldo:

$ sudo nano /usr/share/applications/aldo.desktop

In the resulting editor type the following:

```
[Desktop Entry]
Name=Aldo Morse Code Trainer
Comment=Amateur Radio Morse Code Trainer program
TryExec=aldo
Exec=aldo %F
Icon=aldo.png
Terminal=true
Type=Application
Categories=HamRadio;
```

Aldo Desktop Entry

Save the file by typing "[Ctrl] X", "Y",[Enter]

Restart the menu with:

$ sudo lxpanelctl restart

Configuration:

Choose option 5: setup. Then choose option 1: Keyer Setup. Press the enter key twice after which the various parameters can be set.

After completing setup of the program option 6 will return you to the main menu.

Operating:

Click [Menu] [HamRadio] [Morse Code Training]. Select the type of code you wish to have generated. If a HDMI HD TV is being used as the monitor, the sound will be heard from the monitor's speaker or if the monitor has no speaker the sound can be heard from the monitor's audio output jack.

Chirp: Configuration tool for amateur radios

Overview:

Chirp is a tool for saving, restoring, and managing memory and preset data in amateur radios. It supports a large number of manufacturers and models, as well as provides a way to interface with multiple data sources and formats.[4]

Supported Radio Models can be found at http://chirp. danplanet.com/projects/chirp/wiki/Supported_Radios

Hardware:

This program requires a USB cable to connect with your equipment. See http://chirp.danplanet.com/projects/chirp/wiki/CableGuide for further information.

Installation:

There are two ways to install software. Choose one of the following methods:

Installation by command line:

Open a terminal session by clicking on the terminal icon on the task bar at the top of the screen.

If not recently done update information about new and updated Linux packages that are available:

$ sudo apt-get update

If a new version of a package installed on your machine is available the package will be upgraded with the following command. No installed packages are ever removed by this command.

$ sudo apt-get upgrade

Then in order to install the program:

$ sudo apt-get install chirp

In response to the 'Do You Want to Continue [Y/n]' prompt, press the [enter] key.

When the download is complete you have completed the installation.

Installation by Add/Remove Software menu selection:

Click on the Raspberry Icon on the task bar.

Click on preferences on the drop down list.

Select Add/Remove Software.

Enter chirp in the search box and press enter.

Click on the "Configuration tool for amateur radios" selection square.

Click on the Apply button.

Enter the pi password.

When the download is complete you have completed the installation.

Desktop Entry File:

The Desktop Entry File is automatically generated.

Configuration:

There are no initial configuration requirements.

Operating:

A beginner's guide can be found at http://chirp.danplanet. com/projects/chirp/wiki/Beginners_Guide

Fldigi: Digital Modem Program

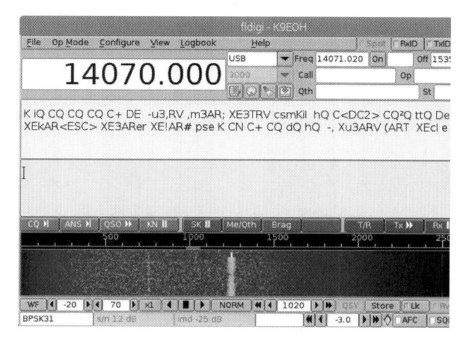

Flidigi modem screen

Overview:

Fldigi is a modem program which supports most of the digital modes used by ham radio operators today. You can also use the program for calibrating your sound card to WWV or doing a frequency measurement test. The program also comes with a CW decoder.

Flarq: By opening Fldigi first, then Flarq, the ability to transmit and receive frames of Automatic Repeat request (ARQ) data results in correction of errors in transmission.

Flmsg is a simple forms management editor for the amateur radio supported standard message formats, including ICS, HICS, MARS, IARU, NTS Radiograms, Red Cross and plain text. Its data files are pure ASCII text that can be sent from point to point using the internet, amateur radio or other electronic links.

Flwrap is a small desktop application that encapsulates a text file, an image file, or a binary file within a set of identifier blocks. Flwrap is designed to be used to best advantage with fldigi but can be used with any digital modem program.

Hardware:

Since the Raspberry Pi has no audio input capability you will need an external audio adapter such as the Tigertronics SignaLink USB adapter

If you possess a Computer Assisted Transceiver (CAT) the fldigi software and the transceiver can pass information back and forth. If the transceiver frequency is changed the display on fldigi will so indicate and can be used to automatically enter the frequency and time information in the built in logging program in the fldigi software.

A cable will be required for the CAT function. The manufacturer probably has a cable available to attach between the transceiver and computer. Check with a distributor of your transceiver to obtain the cable. However, these cables generally use RS-232 specifications. You would therefore also need a RS-232 to USB adapter for the Raspberry Pi connection. Search for "RS-232 to USB adapter" to find one on such sites as amazon.com.

Warning: When used on the air with transmitting equipment be aware the program creates a 100% duty cycle with high modulation levels. Please refer to your transceiver manual and reduce your transmitter output power to a safe level.

Installation:

There are two ways to install software. Choose one of the following methods:

Installation by command line:

Open a terminal session by clicking on the terminal icon on the task bar at the top of the screen.

If not recently done update information about new and updated Linux packages that are available:

$ sudo apt-get update

If a new version of a package installed on your machine is available the package will be upgraded with the following command. No installed packages are ever removed by this command.

$ sudo apt-get upgrade

Press the [Enter] key in response to the 'Do You Want to Continue [Y/n]' prompt.

Then in order to install the program:

$ sudo apt-get install fldigi flmsg flwrap

Press the [Enter] key in response to the 'Do You Want to Continue [Y/n]' prompt.

When the download is complete you have completed the installation.

Installation by Add/Remove Software menu selection

Click on the Raspberry Icon on the task bar.

Click on preferences on the drop down list.

Select Add/Remove Software.

Enter fldigi in the search box and press enter.

Click on the "digital modem program for hamradio operators", "ham radio transceiver control program" and "amateur radio file encapsulation/compression utility" selection squares.

Click on the [Apply] button.

After the above are downloaded, enter flmsg in the search box and press enter.

Click on the "amateur radio forms management editor"

Click on the [Apply] button

If requested, enter the pi password.

When the download is complete you have completed the installation.

Desktop Entry File:

Both Fldigi and Flarq are bundled in this package.

The Desktop Entry Files for both are automatically created.

44

However, in addition to the HamRadio category, Network is included.

Should you wish to remove the Network entry:

$ sudo nano /usr/share/applications/fldigi.desktop

In the resulting editor remove Network; from the Categories line with the following results.

```
[Desktop Entry]
Name=Fldigi
GenericName=Amateur Radio Digital Modem
Comment=Amateur Radio Sound Card Communications
Exec=fldigi
Icon=fldigi
Terminal=false
Type=Application
Categories=HamRadio;
```

Fldigi Desktop Entry

Save the file by "[Ctrl] X", "Y" and [Enter].

Do the same for Desktop Entries:

$ sudo nano /usr/share/applications/flmsg.desktop

$ sudo nano /usr/share/applications/flwrap.desktop

$ sudo nano /usr/share/applications/flarq.desktop

Update the Menus:

$ sudo lxpanelctl restart

Configuration:

Before opening fldigi, connect your audio interface and transceiver control cable if you are using them.

The first time the program is run a 'Fldigi configuration wizard" is displayed. Click on [Next]

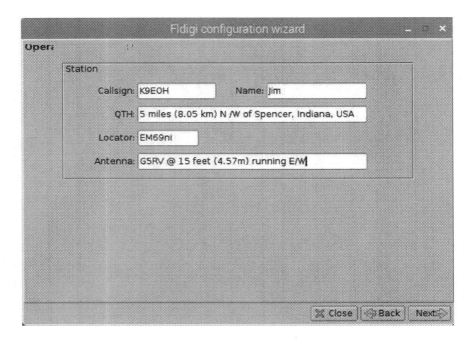

Fldigi Operator configuration wizard

Enter your Callsign, Name, Location (QTH), Maidenhead locator[5] (Locator) and Antenna information. Click on the [Next] key.

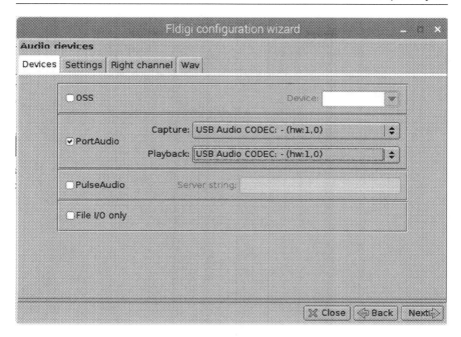

Fldigi Audio devices Configuration

For SignaLink USB interface click on [Devices] tab and check the Port Audio Box. For Capture, select USB Audio CODEC: USB Audio (hw:1,0). For Playback, select USB Audio CODEC: USB Audio (hw:1,0). Select Right channel tab.

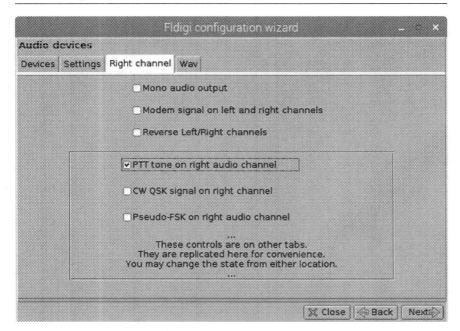

Fldigi Audio devices Right Channel Configuration

Click on the [Right channel] tab. Then select 'PTT tone on right audio channel.' Click on [Next]

In the Transceiver control window click on the Hamlib tab.

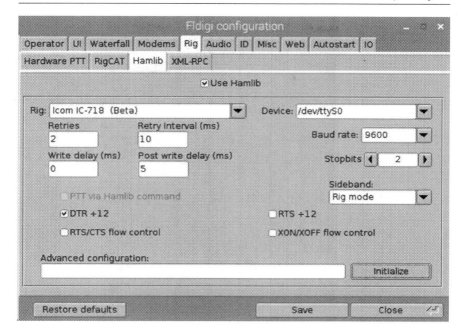

Fldigi Hamlib Configuration

Select the [Hamlib] tab. Select your rig from the drop down menu. Select /dev/ttyS0 from the Device: drop down menu. Set the Baud rate for you rig. You may have to experiment with the four boxes (DTR +12, RTS +12, RTS/CTS flow control and XON/XOFF flow control) to make the Hamlib operate your rig. Click on the [Use Hamlib] check box. Click on the [Initialize] button. Click on [Next] until the program starts.

Click on the Configure selection at the top of the fldigi program window. Click [Misc] on the drop down window.

Click on [CPU], confirm that the Slow CPU box is not checked if using any of the Raspberry Pis listed in Appendix A as all are 700 MHz or faster.

Click the [NBEMS] tab.

If you desire incoming messages to be displayed on a flmsg form check the 'Open with flmsg' check box.

If you wish a copy also be placed in a browser HTML format, click the 'Open in browser' check box.

Open a terminal session by clicking on the terminal icon on the task bar at the top of the screen.

To find the location of flmsg, enter the following:

$ sudo find / -name flmsg

```
pi@HAMSHACK:~ $ sudo find / -name flmsg
/usr/share/doc/flmsg
/usr/bin/flmsg
```

Search for flmsg location

As shown above, enter the line /usr/bin/flmsg (the one without the /doc/ in the flmsg: window) Click [Save] and [Close]

Restart the program by clicking on the [x] in the upper right corner. Confirm Quit by clicking on the [Yes] button.

Then Select fldigi via the menu system [Raspberry Icon] (in the task bar on top of the screen, [Hamradio] and [Fldigi]

Operating:

Digital bands include 1.805-1.838, 3.522-3.620, 7.025-7080, 10,137-10.142, 14.070-14.107, 18.098-18.106, 21.070-21.540, 24.197-24.922 and 28.076-28.120 MHz.

For a complete list including modes go to URL ciarc.org/downloads/Digital_Mode_Band_Plan.pdf

A beginner's Guide is available by clicking Help on the top bar of the program.

gpredict: Satellite Tracking

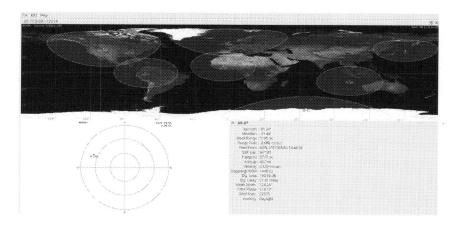

Gpredict screen

Overview:

Gpredict is a real-time satellite tracking and orbit prediction application. It can track an unlimited number of satellites and display their position and other data in lists, tables, maps, and polar plots (radar view). Gpredict can also predict the time of future passes for a satellite, and provide you with detailed information about each pass.

Gpredict is different from other satellite tracking programs in that it allows you to group the satellites into visualization modules. Each of these modules can be configured independently from the others giving you unlimited flexibility concerning the look and feel of the modules. Naturally, Gpredict will also allow you to track satellites relatively to different observer locations - at the same time.

Hardware:

It is possible to provide Doppler tuning of radios and tracking of antenna rotators by using Hamlib. See Section 7 "Controlling Radio and Rotators" at http://www.k9eoh. com/gpredict-user-manual-1_3.pdf for a comprehensive explanation of the process.

Installation:

There are two ways to install software. Choose one of the following methods:

Installation by command line:

Open a terminal session by clicking on the terminal icon on the task bar at the top of the screen.

If not recently done update information about new and updated Linux packages that are available:

$ sudo apt-get update

If a new version of a package installed on your machine is available the package will be upgraded with the following command. No installed packages are ever removed by this command.

$ sudo apt-get upgrade

Press the [Enter] key in response to the 'Do You Want to Continue [Y/n]' prompt.

Then in order to install the program:

$ sudo apt-get install gpredict

When the download is complete you have completed the installation.

Installation by Add/Remove Software menu selection:

Click on the Raspberry Icon on the task bar.

Click on preferences on the drop down list.

Select Add/Remove Software.

Enter "gpredict" in the search box and press enter.

Click on the "Satellite tracking program" selection square.

Click on the Apply button.

If prompted enter the pi password.

When the download is complete you have completed the installation.

Desktop Entry File:

The application automatically creates the Desktop Entry file. However, the categories of Network, HamRadio, Education, Astronomy and Science are all present. This will place a button for the application in these multiple Menu selections.

If you wish to remove the Menu selections you may edit the Desktop Entry file as follows.

Open a terminal session by clicking on the terminal icon on the task bar at the top of the screen.

$ sudo nano /usr/share/applications/gpredict.desktop

This will open the nano editor. On the Categories line, delete all except HamRadio; as shown.

Gpredict Desktop Entry

Save the file by "[Ctrl] X", "Y" and [Enter].

Update the Menu:

$ sudo lxpanelctl restart

GNOME Predict is now listed only in the HamRadio submenu.

Configuration:

After opening the program from the menu, the first step is to update the TLE files by clicking [Edit] [Update TLE] [from Network]. When the update shows Finished, click on [Close] button.

Then, select [Edit] [Preferences]. Select your Number Format preferences. If you are in the US perhaps you will change show local time instead of UTC and from Metric to imperial units.

Click on [Ground Stations] [Add New].

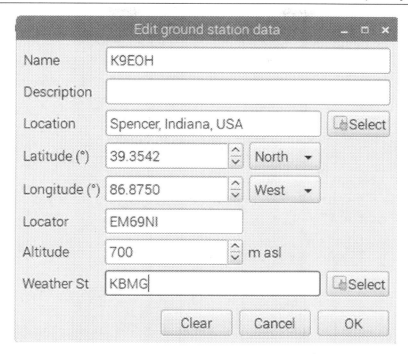

Edit ground station data

Fill out the Edit Ground Stations Data box with your information.

For [Name] I entered my call sign.

In the [Location] box type in the town.

For locator enter your maidenhead locator and the latitude and longitude will be automatically entered.

Enter your altitude above mean sea level in the Altitude window.

For [Weather St], click on [Select] and click through to the nearest town. Click [OK].

Click on the Sample entry and [Delete] to remove it from the program.

Press the [Enter] key in response to the 'Do You Want to Continue [Y/n]' prompt.

Also, on the left of the GPREDICT Preferences is a Modules icon. To select different views click on it to set them up. You may [Select layout] and view the various options with a small window of the layout. After selecting your favorite, click on the [OK] button. When finished, click on [OK].

Restart the program to invoke the changes.

Operating:

A listing of satellite frequencies can be downloaded at http://www.ne.jp/asahi/hamradio/je9pel/satslist.xls

The gpredict user manual for version 1.3 can be downloaded from http://www.k9eoh.com/gpredict-user-manual-1_3.pdf

Trusted QSL-Logbook of the World

Overview:

QSL is a confirmation of contact between two amateur radio stations. The ARRL Logbook of the World project is a database which collects data about contacts between amateur stations (QSOs). This package provides the 'tqsl' program for maintaining your digital certificates for Logbook of The World (LoTW) and signing and uploading QSO log files. Go to http://www.arrl.org/files/file/LoTW%20 Instructions/Quick%20Start/TQSL%202_0.pdf for an overview and tutorial.

Hardware:

There are no additional hardware requirements.

Installation:

There are two ways to install software. Choose one of the following methods:

Installation by command line:

Open a terminal session by clicking on the terminal icon on the task bar at the top of the screen.

If not recently done update information about new and updated Linux packages that are available:

$ sudo apt-get update

If a new version of a package installed on your machine is available the package will be upgraded with the following command. No installed packages are ever removed by this command.

$ sudo apt-get upgrade

Press the [Enter] key in response to the 'Do You Want to Continue [Y/n]' prompt.

Then in order to install the program:

$ sudo apt-get install trustedqsl

Press enter in response to the 'Do You Want to Continue [Y/n]' prompt.

When the download is complete, you have completed the installation.

Installation by Add/Remove Software menu selection:

Click on the Raspberry Icon on the task bar.

Click on preferences on the drop down list.

Select Add/Remove Software.

Enter qsl in the search box and press enter.

Click on the "QSL log signing for the Logbook of the World(LotW)" check box.

Click on the Apply button.

Enter the pi password.

Desktop Entry File: Downloading this software creates the Desktop Entry File.

Configuration:

Steps involved to configure include 1) Request of a certificate for your callsign, 2) Authentication of your location, 3) acceptance of your Call Sign Certificate, 4) creation of a station location and 5) sign and upload of your log files via the internet. Go to http://www.arrl.org/quick-start-tqsl for a quick start.

Operation:

See a list of guides at http://www.arrl.org/quick-start

Gworldclock - World Clock

Overview:

This program displays the time and date of specified time zones using a GTK+ interface. It also provides a "rendezvous" function allowing the zones to be synchronized to a time other than the current time. This can help you organize meetings across different timezones.

The time zones may be entered by hand in time zone format or chosen from a list prepared from /usr/share/zoneinfo/ zone.tab. The zone list is kept in a format consistent with the shell script tzwatch.

Hardware:

There are no additional hardware requirements.

Installation:

There are two ways to install software. Choose one of the following methods:

Installation by command line:

Open a terminal session by clicking on the terminal icon on the task bar at the top of the screen.

If not recently done update information about new and updated Linux packages that are available:

$ sudo apt-get update

If a new version of a package installed on your machine is available the package will be upgraded with the following command. No installed packages are ever removed by this command.

$ sudo apt-get upgrade

Then in order to install the program:

$ sudo apt-get install gworldclock

Installation by Add/Remove Software menu selection:

Click on the Raspberry Icon on the task bar.

Click on preferences on the drop down list.

Select Add/Remove Software.

Enter gworldclock in the search box and press enter.

Click on the "Displays time and date in specified time zones" check box.

Click on the Apply button.

Enter the pi password.

Desktop Entry file:

Although automatically created includes categories other than HamRadio. If you wish to and Ham Radio and remove the program from the other menus:

$ sudo nano /usr/share/applications/gworldclock.desktop

This will open the Desktop Entry file. Down arrow to the Categories entry and replace all Category entries with HamRadio; as shown.

```
[Desktop Entry]
Comment=See the time in other timezones
Name=World Clock (gworldclock)
Encoding=UTF-8
Comment[fr]=Voir l'heure dans d'autres fuseaux horaires
Exec=gworldclock
Terminal=false
Type=Application
Icon=gworldclock
Categories=HamRadio;
```

GWorldclock Desktop Entry

Save the file by "Ctrl X", "Y" and Enter.

Update the Menu:

$ sudo lxpanelctl restart

Configuration:

To add time zones in a specific city, click on options and "Add Timezone". Click the Continents tab and select a continent. Click on Countries and select a country. Click on Regions and select a city. Click on the [Add Zone] button and the [Done] button.

To add UTC GMT Zulu or Navajo, click on Options and "Add Timezone". Under the Continents tab select "Universal". Click on the Regions tab and select UTC, GMT, Zulu or Navajo. Click on the [Add Zone] button and the [Done] button.

Operating:

The clock will display all time zones selected.

Xlog: a logging program for Hams

Overview:

Xlog is a logging program for amateur radio operators which can be used for daily logging and contests. Logs are stored into a text file. QSO's are presented in a list. Items in the list can be added, deleted or updated. For each contact, DXCC information is displayed and bearings and distance is calculated, both short and long path.

When Hamlib is enabled through the menu, you can retrieve frequency, mode and signal-strength from your rig over the serial port.

Hardware:

If you possess a Computer Assisted Transceiver (CAT), using the cable described in the next paragraph and enabling Hamlib as described in the Configuration section, the frequency, mode and signal-strength can be automatically loaded in the xfile database.

The manufacturer probably has a cable available to attach between the transceiver and computer. However, these cables generally use RS-232 specifications. You would therefore also need a RS-232 to USB adapter for the Raspberry Pi connection. Search for "RS-232 to USB adapter" to find one on such sites as amazon.com.

Installation:

There are two ways to install software. Choose one of the following methods:

Installation by command line:

Open a terminal session by clicking on the terminal icon on the task bar at the top of the screen.

If not recently done update information about new and updated Linux packages that are available:

$ sudo apt-get update

If a new version of a package installed on your machine is available the package will be upgraded with the following command. No installed packages are ever removed by this command.

$ sudo apt-get upgrade

Press the [Enter] key in response to the 'Do You Want to Continue [Y/n]' prompt.

Then in order to install the program:

$ sudo apt-get install xlog

Press [enter] in response to the 'Do You Want to Continue [Y/n]' prompt.

When the download is complete you have completed the installation.

Installation by Add/Remove Software:

Click on the Raspberry Icon on the task bar.

Click on preferences on the drop down list.

Select Add/Remove Software.

Enter xlog in the search box and press enter.

Click on the "GTK+Logging program for Hamradio Operators" check box.

Clock on the "data for xlog, a GTK+lLogging program for Hamradio Operators"

Click on the Apply button.

Enter the pi password.

Desktop Entry File:

Although automatically created, the desktop entry file includes categories in addition to HamRadio. If you wish to remove the program from the other sub-menus:

$ sudo nano /usr/share/applications/xlog.desktop

This will open the Desktop Entry file.

On the Categories line, delete all except HamRadio; as shown.

```
[Desktop Entry]
Name=Xlog
Comment=Amateur Radio logging program
Comment[nl]=Log programma voor zendamateurs
Comment[pl]=Dziennik łączności dla radioamatorów
TryExec=xlog
Exec=xlog %F
Icon=xlog-icon.png
Terminal=false
Type=Application
Categories=HamRadio;
MimeType=text/x-xlog
```

Xlog Desktop Entry

Save the file by "[Ctrl] X", "Y" and [Enter].

Update the Menu:

$ sudo lxpanelctl restart

Configuration:

Open the program and click on [OK] in the first xlog-setup windows. Then click on [Settings] [Preferences] [Info]. Enter your call sign. By entering your maidenhead coordinate in the QTH locator box, the Latitude and Longitude boxes will be filled in. Set the Units 'distance in' to your preference of Kilometers or Miles.

If you have the cable described in the hardware section, click on the [Hamlib] tab.

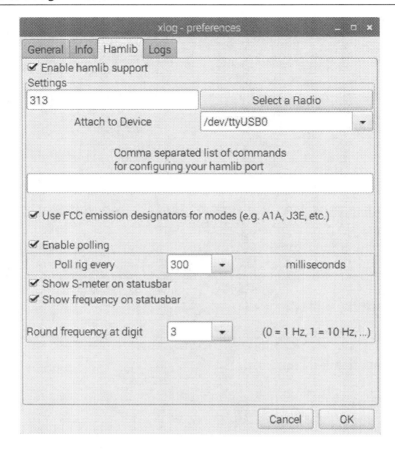

Xlog preferences tab Hamlib

Click on the [Enable hamlib support] check box.

Select your radio model and click on the [OK] button.

Click on the [Use FCC emission designators for modes], [Enable polling], [Show S-meter on statusbar] and [Show frequency on statusbar] check boxes.

In order to determine the Device address, open a terminal and enter:

$ ls /dev/tty*

Select from the 'Attach to Device' device address that matches one of those displayed with the terminal listing.

Click on the [OK] button.

The s-meter and frequency information should be displayed at the bottom of the xlog main window. If not, the 'Attach to Device' entry probably needs to be changed.

Operating:

For more details, select [Help][Manual] or Ctl+H.

Xnec2c: Calculate and Display Antenna Properties

Overview:

GTK based graphical wrapper for nec2c. The program incorporates nec2c and is interactive, presenting results graphically, as they are calculated.

This includes color-coded radiation patterns, current/charge distribution and graphs of gain, impedance etc. A basic NEC2 input file editor is now built-in.

Hardware:

There are no additional hardware requirements.

Installation:

There are two ways to install software. Choose one of the following methods:

Installation by command line:

Open a terminal session by clicking on the terminal icon on the task bar at the top of the screen.

If not recently done update information about new and updated Linux packages that are available:

$ sudo apt-get update

If a new version of a package installed on your machine is available the package will be upgraded with the following command. No installed packages are ever removed by this command.

$ sudo apt-get upgrade

Press the [Enter] key in response to the 'Do You Want to Continue [Y/n]' prompt.

Then in order to install the program:

$ sudo apt-get install xnec2c

When the download is complete you have completed the installation.

Installation by Add/Remove Software:

Click on the Raspberry Icon on the task bar.

Click on preferences on the drop down list.

Select Add/Remove Software.

Enter xnec2c in the search box and press enter.

Click on the [GTK+Logging program for Hamradio Operators] check box.

Click on the [data for xlog, a GTK+lLogging program for Hamradio Operators]

Click on the Apply button.

Enter the pi password.

Desktop Entry File:

The Desktop Entry File is automatically created. However, there is no HamRadio category listed. In order for the Xnex2c to appear in the HamRadio menu changes will need to be made to the Categories listing.

In order to do so:

$ sudo nano /usr/share/applications/xnec2c.desktop

In the resulting editor change the Categories line to:

```
[Desktop Entry]
Name=Xnec2c
Comment=Xnec2c antenna simulation
Exec=xnec2c
Terminal=false
Type=Application
Categories=HamRadio:
Keywords=antenna;simulation;
```

Xnec2 Desktop Entry

Save the file by "[Ctrl] X", "Y" and [Enter].

Update the Menu:

$ sudo lxpanelctl restart

Configuration:

None required.

Operating:

A manual can be found at http://www.qsl.net/5b4az/pkg/nec2/xnec2c/doc/xnec2c.html

Examples can be found at /usr/share/doc/xnec2c/examples which can be used to learn the operation of Xnex2c.When you open the program, click on file, Open, File System, usr, share, doc, xnec2c, examples and a list of nec files will be displayed. Click on any of interest.

APPENDIX A

Current Pi Model Information

Raspberry Pi Model A+

The Model A+ is the low-cost variant of the Raspberry Pi. It replaced the original Model A in November 2014.

The GPIO header has 40 pins, while retaining the same pinout for the first 26 pins as the Model A and B.

The old friction-fit micro SD card socket has been replaced with a much nicer push-push micro SD version.

The linear power regulator with switching uses only 0.5W to 1W.

The audio circuit incorporates a dedicated low-noise power supply.

Form factor. The USB connector is aligned with the board edge, composite video is available on the 3.5mm jack, and there are four squarely-placed mounting holes. Model A+ is approximately 2cm shorter than the Model A.

The Model A+ is recommended for embedded projects and projects which require very low power, and which do not require Ethernet or multiple USB ports.

Raspberry Pi Zero

The Raspberry Pi Zero is half the size of a Model A+, with twice the utility. A tiny Raspberry Pi that's affordable enough for any project!

1Ghz, Single-core CPU

512MB RAM

Mini HDMI and USB On-The-Go ports

Micro USB power

HAT-compatible 40-pin header

Composite video and reset headers

Raspberry Pi 2 Model B

The Raspberry Pi 2 Model B is the second generation Raspberry Pi. It replaced the original Raspberry Pi 1 Model B+ in February 2015. The Raspberry Pi 1 has:

A 900MHz quad-core ARM Cortex-A7 CPU

1GB RAM

Like the (Pi 1) Model B+, it also has:

4 USB ports

40 GPIO pins

Full HDMI port

Ethernet port

Combined 3.5mm audio jack and composite video

Camera interface (CSI)

Display interface (DSI)

Micro SD card slot

VideoCore IV 3D graphics core

Because it has an ARMv7 processor, it can run the full range of ARM GNU/Linux distributions, including Snappy Ubuntu Core, as well as Microsoft Windows 10.

Raspberry Pi 3 Model 3

The Raspberry Pi 3 is the third generation Raspberry Pi. It replaced the Raspberry Pi 2 Model B in February 2016. It has:

A 1.2GHz 64-bit quad-core ARMv8 CPU

802.11n Wireless LAN

Full HDMI port

Ethernet port

Combined 3.5mm audio jack and composite video

Camera interface (CSI)

Display interface (DSI)

Micro SD card slot (now push-pull rather than push-push)

VideoCore IV 3D graphics core

The Raspberry Pi 3 has an identical form factor to the previous Pi 2 (and Pi 1 Model B+) and has complete compatibility with Raspberry Pi 1 and 2.

The Raspberry Pi 3 Model B is recommended for use in schools, or for any general use. Those wishing to embed their Pi in a project may prefer the Pi Zero or Model A+, which are more useful for embedded projects, and projects which require very low power.

APPENDIX B

CLI Useful commands

The **$** indicates the terminal prompt.

Configure: $ sudo raspi-config

Restart menu: $ sudo lxpanelctl restart

If changes have been made to a Desktop Entry file, the menu should be restarted to view the change in the menu system.

Synchronize repository file list: $ sudo apt-get update

Running sudo apt-get update makes sure your list of packages from all repositories is up to date.

Update installed files: $ sudo apt-get upgrade

Installs newest versions of all packages currently installed on the system. If a new version of a package installed on your machine is available the package will be upgraded with this command. No installed packages are ever removed by this command.

Program installation: $ sudo apt-get install [program]

Program removal: $ sudo apt-get remove [program]

Update Raspbian firmware: $ sudo apt-get install rpi-update

Super user: **$** sudo

Be careful using super user mode. This mode allows system administration with access to all areas of the operating system and use of all commands, some of which can be destructive if improperly applied.

System shutdown: sudo shutdown –h now

System reboot: sudo reboot –h now

APPENDIX C

Useful websites

Raspberry Pi official site: https://www.raspberrypi.org

Download Raspberry Pi system images: https://www.raspberrypi.org/downloads/

Raspberry Pi Connect: http://www.raspberryconnect.com/

Raspberry Connect is a site for the Raspberry Pi community to post articles about their Raspberry Pi projects and activities or promote Raspberry Pi websites in a web directory.

Adafruit learning page: https://learn.adafruit.com/adafruit-raspberry-pi-lesson-1-preparing-and-sd-card-for-your-raspberry-pi

This is first in a set of 13 lessons for learning to use the Raspberry Pi.

APPENDIX D

Bibliography

"Computers in Amateur Radio" Steve White G3ZVW Radio Society of Great Britain Publication.

"Get on the Air with HF Digital" Steve Ford WB9IMY ARRL Publication

"Raspberry Pi Hacks" Ruth Suehle Tom Callaway O'Reilly Media, Inc. Publication.

"Raspberry Pi User Guide" Eben Upton and Gareth Halfacree John Wiley & Sons Ltd. Publication.

"Computer Lib/Dream Machines" by Ted Nelson, self published, now available in reprint for about $100. Used editions are listed for as much as $1000.

"Raspberry Pi Projects for the Evil Genius" by Donald Norris McGraw_Hill Education.

APPENDIX E

Glossary

ASCII: American Standard Code for Information interchange the most common format for text files in computers and on the internet.

Boot: Booting process initially loads an operating system into the computer.

CAT: Computer Aided Transceiver is a radio that has the capability of interfacing with external programs.

CLI: Command Line Interface

CPU: Central Processing Unit: that portion of the computer that interprets and executes program instructions.

DXCC: The DX Century Club is an amateur radio award given to those amateurs contacting at least 100 countries. The program xlog can keep track of the number of countries contacted toward the DXCC award.

GPIO: General Purpose Input Output: Pins on the pi that can sense and control voltages to remote entities such as temperature sensors and voltage switching relays.

GPU: Graphics Processing Unit: portion of the computer what accelerates operation of the screen graphics.

GUI: Graphic User Interface uses windows, icons and menus which can be manipulated by a mouse.

HDMI: A high Definition Multimedia Interface is a standard ensuring that cables and connectors properly transfer high definition signals.

HAT: Hardware Attached on Top. A board that fits on the

top of a Raspberry Pi containing circuits that extend the functions of the Raspberry Pi.

HICS: Hospital incident command system is designed for hospitals for use in emergency and non-emergency situations. A message form used by the HICS is included in the flmsg program.

IARU: The International Amateur Radio Union represents amateur radio societies around the world. A message form used by the IARU is included in the flmsg program.

ICS: a standardized on-scene incident management concept designed specifically to allow responders to adopt an integrated organizational structure equal to the complexity and demands of any single incident or multiple incidents without being hindered by jurisdictional boundaries. A message form used by the ICS is included in the flmsg program.

IP address: is a unique string of numbers separated by periods that identifies each computer using the internet.

Maidenhead locator: is a robust geographic coordinate system to describe amateur radio station locations.

MARS: A US Department of Defense sponsored program managed and operated by the US Army and US Air Force consisting primarily of amateur radio operators. MARS message forms are included in the flmsg program.

Microprocessor: A single chip integrated circuit CPU.

Modem: modulator demodulator is a device to convert data to carrier wave signals on the internet or telephone phone lines.

NOOBS: New Out Of Box Software used to initially boot up the Raspberry Pi and providing different operating system choices.

NTS: Amateur radio National Traffic System is designed to communicate information critical to savings lives or property during disasters. The standard form used on the National Traffic System is the Radiogram. This form is included in the flmsg program.

Open Source Software: is licensed to allow software to be freely used, modified and shared.

OS: Operating System is the program that operates all other programs in a computer.

Repository: Repositories are servers which contain sets of software packages.

Tarball: is a method of compressing and decompressing files that are bundled together in order to decrease the time required to load a program and to save space on disk drives.

Task Bar: The bar generally at the top of the desktop screen containing icons for starting programs, showing what programs or applications are running. It also provides links to other places and programs.

APPENDIX F

End Notes

1   SETI – Search for ExtraTerrestrial Intelligence https://www.seti.org/node/647.

2   Raspberry Pi Projects for the Evil Genius, Donald Norris and Home Automation with the Raspberry Pi, Marco Schwartz for examples.

3   Adding a printer to raspberry pi http://www.howtogeek.com/169679/how-to-add-a-printer-to-your-rasperry-pi-or-other-Linux-computer

4   Chirp – chirp.danplanet.com/projects/chirp/wiki/Home

5   Maidenhead Locator System www.levinecentral.com/ham/grid_square.php

Printed in the United States
By Bookmasters